Contents

What is a species?

As soon as you start to become interested in living things, you will see that there is an enormous variety. However, among the different kinds of living things there are some definite patterns. When you see a group of animals or plants which are very similar to one another, you may say that these animals or plants belong to one **species**.

A species is a group of individuals which can all **interbreed** to produce offspring that can breed – they are **fertile**.

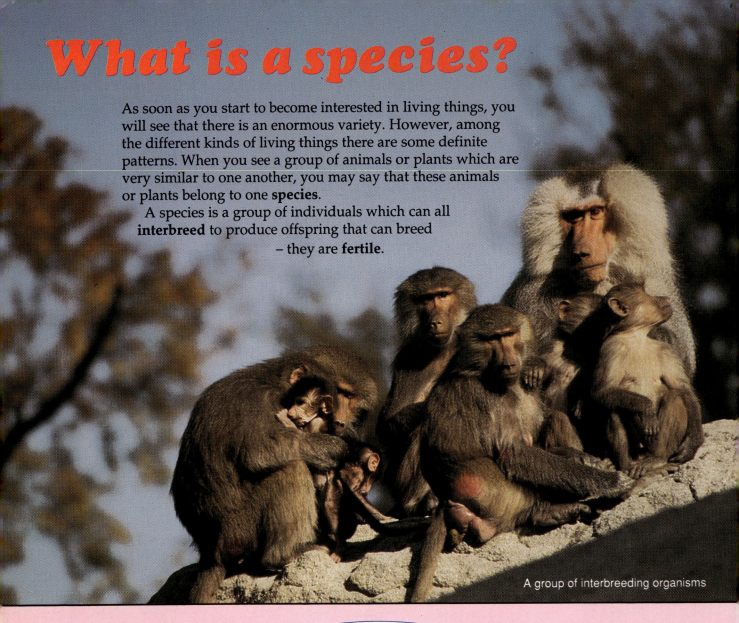

A group of interbreeding organisms

How we classify species

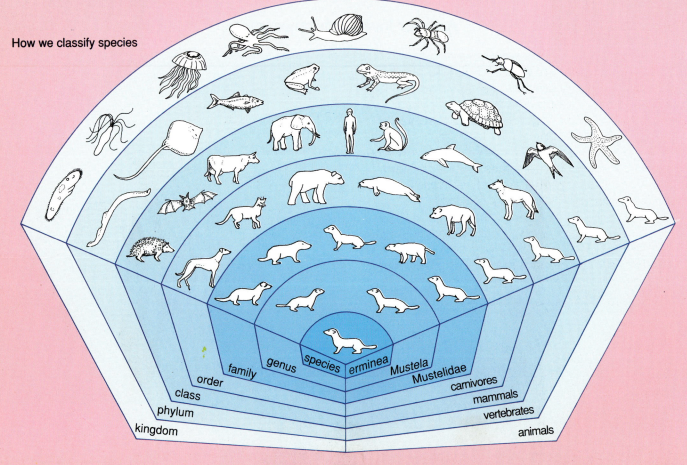

kingdom
phylum
class
order
family genus species erminea Mustela
Mustelidae
carnivores
mammals
vertebrates
animals

Members of a species, besides being similar to one another, must breed together to keep the species going. Animals or plants which belong to different species will not, as a rule, breed together. If they do, the offspring are unlike the parents. At zoos lions have occasionally been mated with tigers. The offspring, called **tigrons**, are not very similar to either parent.

△ A mule

Similarly, if a horse breeds with a donkey, the offspring produced is a **mule**, an animal which is unlike either a horse or a donkey. In these cases the offspring – either mule or tigron – are not able to breed. They are **sterile**.

◁ A tigron

NOW TRY THESE!

The **Crustacea, Insecta** and **Arachnida** are three **classes** of the **phylum Arthropoda**. Although the three classes all belong to the same phylum, crustaceans differ from insects and arachnids by having a large number of jointed legs and two pairs of antennae. The classes of the phylum are further sub-divided into **orders**, and orders into **genera** (plural of **genus**). The order **Diptera** includes insects with one pair of wings.

The housefly has one pair of wings and belongs to the genus *Musca*. It lives in populations which can interbreed to produce fertile offspring because they are all of the species *domestica*.

Classify the housefly by putting the correct terms at **b** to **h**

Kingdom: **a** Animalia
b...:   **c**...
d...:   Insecta
Order: **e**...
f...:   **g**...
Species: **h**...

Housefly

CAN MEMBERS OF THE SAME SPECIES BE DIFFERENT?

All of these dogs feeding at the same bowl belong to the same species. In theory then, they could all interbreed and produce fertile offspring. However, there are very obvious differences in the shapes, sizes and colours. They also differ in temperament and behaviour. How have these different **varieties** developed? Humans wanted dogs with certain characteristics, which they recognized as **mutations** in litters of puppies. They mated those dogs showing the mutations to produce more puppies with the same mutation. After many generations of such **breeding**, new varieties were developed. Remember, though, they are all of the same species.

Variation within a species

Queen Elizabeth II, age 21

Princess Anne at the same age

You can see that, while living things in groups – called species – resemble one another, there are always some differences that distinguish them. The difference in appearance, for example, between a mother and her children is partly due to their different ages and, in the case of sons, different sex. But no woman at 21 is identical to her mother at that age, as a glance through a family album will show.

— Did You Know? —

There is always some variation. Variation is smallest between closely related people. A man resembles his father more closely than he resembles someone else's father. The resemblance between one organism and another becomes less as the relationship between them becomes more distant. When a Scots person is compared to her or his parents or brothers and sisters, there is a greater resemblance than there is if the person is compared with another Scot. But the person resembles all Scots people more closely than Africans or Asians.

Indian

An enormous number of **genetical factors** control the external features which distinguish the members of the different races shown here. It is very difficult to trace racial origins by genetic make-up, because common features do not necessarily indicate the inheritance of certain genes from a common ancestor.

Main physical characteristics, however, can largely be traced back to the adaptation of humans to climate at the dawn of human history. The **genes** for flat cheeks, small noses and mouths which characterize eastern Asian people, for example, are believed to have built up when protection was needed against the extreme cold of the Ice Age.

South African bushman

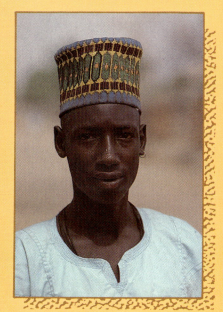

Bororo man, Cameroon (West Africa)

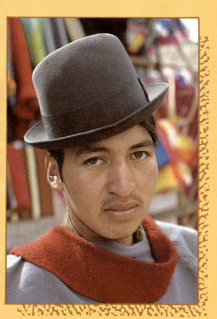

South American Indian

Thai

People of different nationalities

Russian

SURPRISING

species

> you won't catch me going up in one of those!....

Two individuals may look totally different but still be the same species! A **tadpole** is the same species as the **frog** into which it will change. A **caterpillar** is still the same species as the **butterfly** into which it will change.

These change to these – they are the same species!

Tadpoles

Frog

Crab larvae

Adult crab

Starfish larvae

Adult starfish

— Now try these! —

c **1** Two hundred limpets of the same species were collected from a rocky shore. Their shell diameters were measured and then grouped by size in 5 mm ranges. The results are shown in the histogram.

   a What is the most common range?

   b How many limpets measured more than 30 mm in diameter?

   c What percentage of the animals had shells less than 21 mm in diameter?

   d If this species of limpet grows 5 mm each year, how old were the largest limpets in the sample?

2 A gardener wanted to find out the frequency of three flower colours in a mixed bed of Livingstone daisies. Random samples were counted where they grew. The results are shown here.

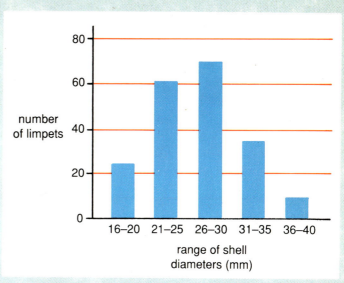

Measuring the shells of limpets – a histogram of results

Sample	Red	Pink	Orange
1 (53 plants)	16	25	12
2 (54 plants)	14	30	10
3 (50 plants)	11	15	24
4 (43 plants)	18	10	15
Total (200)	59	80	61

   a Draw a suitable bar graph to show the total numbers of flowers of each colour.

   b Suggest why more than one sample was taken.

3 The drawings below show **F1** and **F2** generations produced as a result of crossing two varieties of pure-bred dogs.

   a Which two pairs of observable characters (**phenotypes**) are present in the parental generation and subsequent generations?

   b State the dominant forms of these phenotypes observable in the F1.

   c Give a simple explanation why the F2 shows more variation than the members of the F1 do.

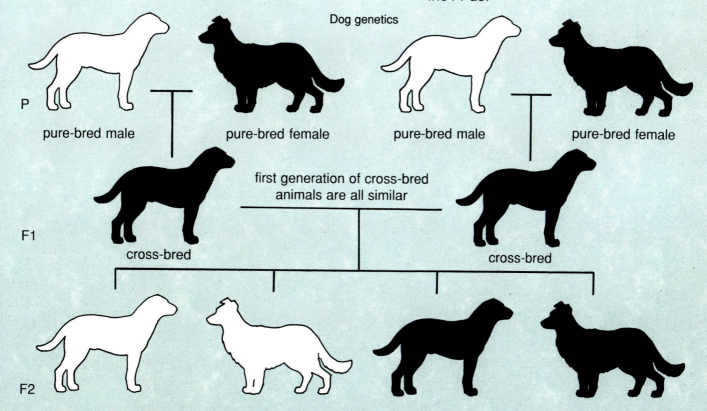

Dog genetics

P — pure-bred male | pure-bred female | pure-bred male | pure-bred female

F1 — cross-bred | first generation of cross-bred animals are all similar | cross-bred

F2

Crossing of two animals from the first cross-bred generation produces a wide variety of offspring

△ Hair comes in different colours

When you look down on a large number of people from a higher level, e.g. a balcony, a variety of hair colour can be seen, from jet black through various shades of brown, grey and red, to white. This type of variation would be impossible to measure using a scale and units. It is caused by many **inherited** differences between the individuals.

The individuals would also differ with regard to their **height** and **weight**. These differences could be measured, and would show a type of variation called **continuous**.

Continuous variations show an even graduation within populations. In a population, the height of each adult could be measured and a histogram constructed of the number of individuals whose heights fall within a given range. Each range should be equal. The histogram would show a curve of **normal distribution.**

Many genes contribute to how tall someone is, for example, those controlling the production of growth hormones and those controlling the rate of protein production. The **environmental factors** involved may be equally diverse, for example, the effects of starvation, malnutrition, disease and lack of exercise.

Discontinuous variation within populations can be seen where individuals fall into two or more distinct groups with respect to one character. It usually occurs when there is a completely **dominant** and a **recessive** character within a pair of contrasted genes. Height in pea plants, investigated by Gregor Mendel (see page 16) is an example of discontinuous variation. This particular plant character is controlled by a single pair of genes and the environment has very little overall effect on it within a large population.

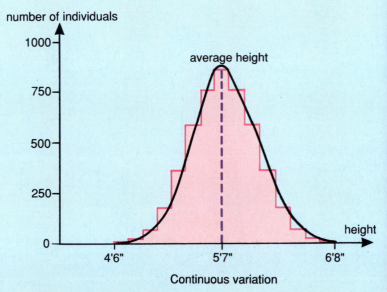

Continuous variation

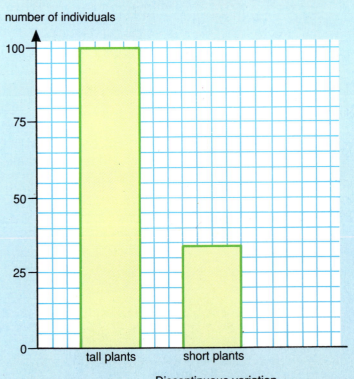

Discontinuous variation

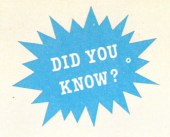

1 The histograms below show variations in two characteristics in a population of goldfish. The population contained 250 fish. Histogram A shows variation in **length**. Histogram B shows variation in **colour**.

a What type of variation is shown by each histogram?

b From histogram B, calculate the number of red/orange goldfish present in the sample.

now try these!

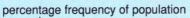

percentage frequency of population

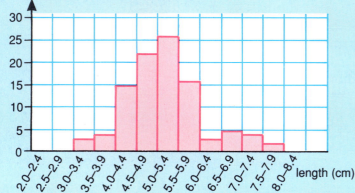

histogram A

percentage frequency in population

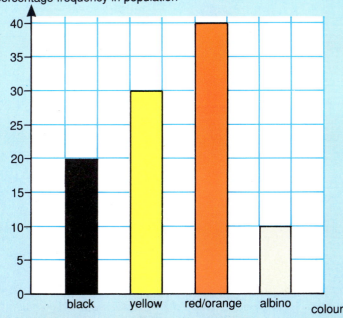

histogram B

2 Two girls studying a small wood near their school decided to measure the thickness of some leaves on a lime tree in the middle of the wood. They did their survey four weeks after the tree came into leaf and took just five similar-sized leaves from three different heights above the ground. Some of their results are shown in the table.

height above ground in m	thickness in mm				
	leaf				
	1	2	3	4	5
2.25	0.8	0.9	0.8	0.8	0.7
6.0	0.5	0.7	0.5	0.6	0.7
8.5	0.3	0.3	0.2	0.3	0.4

Table showing thickness of leaves

a What is the most common thickness of leaves at the 8.5 metre mark?

b Calculate the average thickness of leaves at a height of 2.25 metres.

c Calculate the difference in thickness between the thinnest and thickest leaves found.

d Explain why this variation is more likely to be due to the effects of the environment than to genetical factors.

e Explain how you could test this hypothesis.

f The girls then used a lamp and a light meter to see how much light came through some of the leaves. They found some leaves let 80% of the light through whilst others let only 15% through.
  i State which leaves were found higher up the tree.
  ii Suggest the importance of these results to the tree.

What is inheritance?

Odd one out?

Genetics is the study of the ways in which parents **pass on** characteristics to their offspring. For hundreds of years, farmers knew if they bred from the cows that gave most milk, or from the wheat with the largest grains, they were likely to get these useful features again.

The results of such breeding, however, did not always turn out just as expected. Sometimes the offspring had useful features; at other times they did not. Useful features which did not appear in the children sometimes reappeared in grandchildren. This was a mystery until 1865 when an Austrian monk, **Gregor Mendel**, provided the first scientific explanation for the way features are inherited.

He published the results of his experiments in a local journal of the Brunn Scientific Society where they lay undiscovered until 1900. Since 1900, thousands of scientists have studied genetics using many different species. Certainly not all the answers to genetical problems are yet known. Indeed, the genetical details of few species are known completely.

Gregor Mendel

Some of the best studied organisms are mice, corn, tomatoes, some bacteria, and the fruit fly, *Drosophila*. These have short generations so it has been possible to study **hereditable** features over thousands of generations.

These studies all began with Gregor Mendel and the common garden pea. Mendel kept a garden where he carried out a series of experiments with garden peas. In one series of experiments, Mendel selected plants with **axial** flowers, which always produced plants with axial flowers. He then mated a plant with axial flowers with a plant with **terminal** flowers. All the offspring of this 'cross' had axial flowers. Then he mated two of these cross-bred plants and grew the offspring from the seeds. Instead of all having axial flowers, some of these plants had terminal ones – about one with terminal flowers to every three with axial flowers.

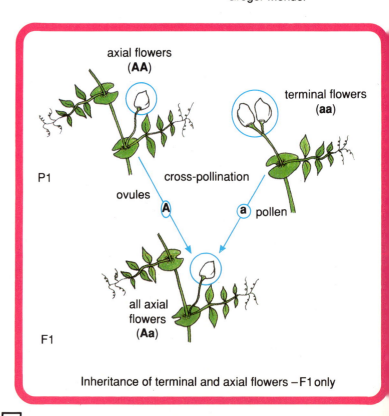

Inheritance of terminal and axial flowers – F1 only

Mendel explained his results by assuming that in a plant each character is controlled by a pair of **hereditary factors**. For instance, flower position is controlled by factors for axial flower and terminal flower. The two factors of a pair separate when sex cells are formed, and each sex cell receives only one factor of the pair. The factors are united in pairs again when two sex cells join to form a new individual.

When Mendel crossed pure-bred plants with axial flowers with pure-bred plants with terminal flowers he got a first generation of all plants with axial flowers. These plants had both factors – one for axial flowers and one for terminal flowers. So, Mendel concluded, the axial factor was **dominant** over the **recessive** terminal factor. When these plants were self-fertilized, the factors segregated again and Mendel got both pure-bred plants with axial flowers and plants with terminal flowers like the ones he had started with, *and* plants with axial flowers which did not breed true.

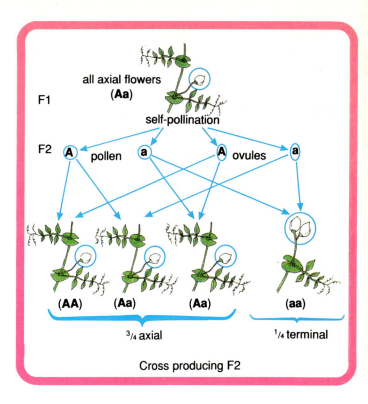

Cross producing F2

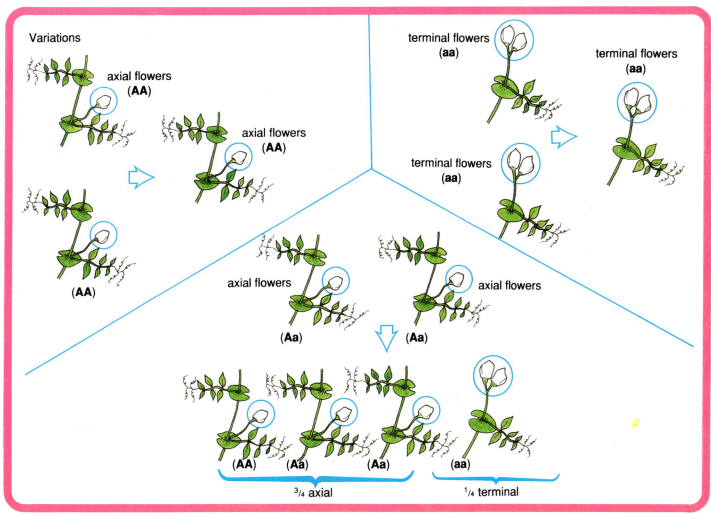

Out of Mendel's experiments came the **Mendelian laws**. These laws can be applied not only to garden peas, but to the inheritance of certain characteristics in most species. By the twentieth century, Mendel's hereditary factors became known as **genes**.

Mendel's Good Luck

It was fortunate that Mendel had the good luck to choose characteristics of peas which were determined by a single pair of genes. In many other organisms a single characteristic may be determined by several genes working together.

The effects of genes may also **blend** so that one of a pair may not show dominance over the other. This can be seen in the blending of flower colour in many species. It is called **incomplete dominance**.

In some flowers, the **allele R** might be supposed to make an enzyme for the development of red pigment, while the allele **W** does not. The **homozygote**, **RR**, makes enough enzyme for full red colour. The homozygote, **WW**, makes no enzyme and is, therefore, white. The **heterozygote**, **RW**, makes some enzyme and is pink.

An example of **codominance** in animals is seen in the case of roan cattle. If a pure-bred red Longhorn cow is mated to a pure-bred white Longhorn bull, the calves are all intermediate roans.

Here, the allele **R** causes red hairs to be produced and the allele **W** causes white hairs. The heterozygote, **RW** has a mixture of red and white hairs in its coat, causing roan colouring.

Another piece of good luck was that the garden pea self pollinates. This means that outside genetical influences, i.e. foreign genes, could not be introduced to experimental plants by cross-pollinating agents such as insects or wind.

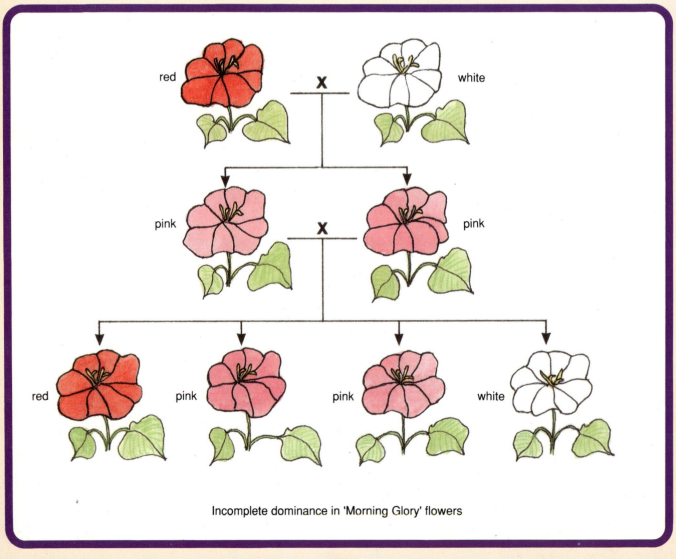

Incomplete dominance in 'Morning Glory' flowers

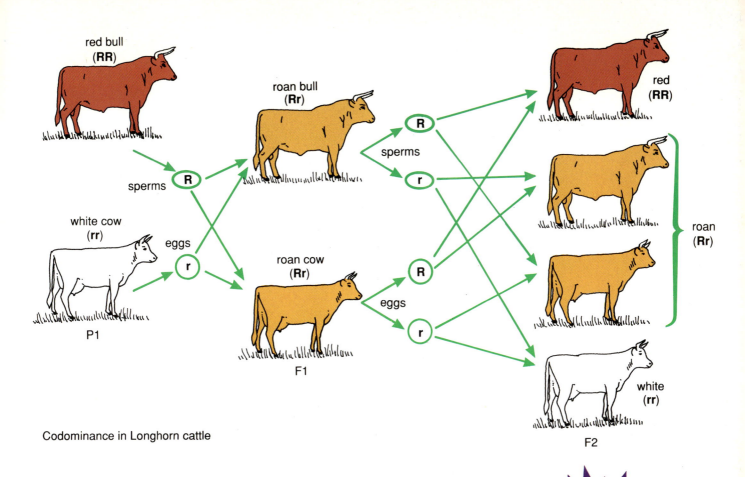

Codominance in Longhorn cattle

Gene Controls Hamster's Clock

In nature, 24-hour clocks are common. Many animals show daily **cycles** of behaviour, being most active at one particular time. These cycles may be partly due to daily changes in light and dark but many cycles continue even when animals are kept in constant light or darkness. Golden hamsters have cycles of activity of about 24 hours when they live in total darkness.

Scientists at the University of Oregon noticed that a male hamster had an abnormally short cycle when it was in total darkness. Even when it was exposed to 14 hours of light and 10 hours of darkness the hamster's period of activity began four hours earlier than usual.

The scientists mated the abnormal hamster with females that had 24-hour cycles. By interbreeding the F1 generation, three groups of hamsters were identified in the F2. One group had a normal 24-hour cycle, another group had a 22-hour cycle and the third group had a 20-hour cycle.

A **mutant** gene seems to be responsible for the short cycle. The scientists suggested that these results are the first evidence that genetic mutations can affect natural rhythms in vertebrates.

groan RING!!

Don't be alarmed!

Now try these!

c 1 Explain what is meant, in this case, by 'mutant gene'.

e 2 The 24-hour cycle is controlled by an incompletely dominant gene. The 22-hour cycle takes place in the heterozygote. Explain how the F2 generation was made up of hamsters with 24-hour cycles, 22-hour cycles and 20-hour cycles in the ratio of 1 : 2 : 1.

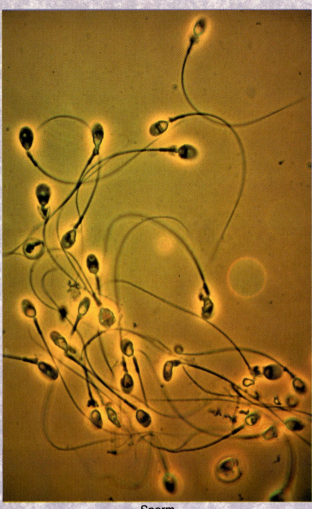

Sperm

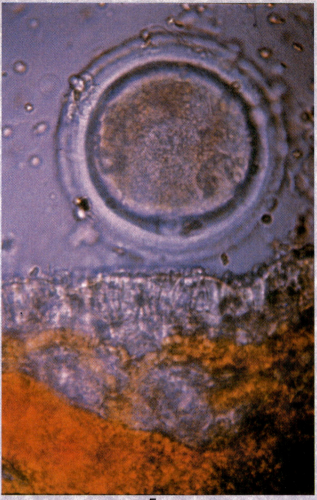

Egg

The largest cells are **eggs**. They contain little except cytoplasm; their bulk is mostly water and stored food. The important part of the egg is its **nucleus**.

Sperm cells are hundreds or even thousands of times smaller than eggs. A sperm cell is little else but a nucleus attached to a tail. When the sperm swims to an egg, its nucleus joins with the egg's nucleus to form the nucleus of a new and separate cell.

The nuclei of both kinds of sex cell generally have fewer **chromosomes** than other cells in the body of the same organism. The reason for this becomes clear when we see what would happen if they did not. Human body cells have 46 chromosomes. If each sex cell, or **gamete**, also had 46 chromosomes, then a baby would have 92, and its children would have 184. Yet all normal cells in human bodies have 46. Human cells normally contain 23 **pairs** of chromosomes. Cell **nuclear division** by a process called **mitosis** ensures that each new cell gets a full set of chromosome pairs.

Sperms and eggs each have only 23 chromosomes, one from each pair. How does an organism produce cells with half the number of chromosome pairs? Of all the millions of cells in

our bodies, only cells in the **testes** and **ovaries** can divide in a way to split up the chromosome pairs. The process is called **reduction division** or **meiosis**. To see how it occurs we can follow meiosis in an insect that has only four pairs of chromosomes.

The first step is similar in some ways to mitosis. Each chromosome pairs off with its opposite number across the middle of the nucleus, and each chromosome duplicates itself. There is a double chromosome for each original, the halves of which are called **chromatids**. Now, just as in mitosis, the nuclear membrane disappears, **but** in this case, each double member of a pair goes to the new cells. At this point we have two cells, each containing four double chromosomes – one from each original pair.

A brief resting period follows, then a new wave of nuclear events begins. Double chromosomes break apart and each chromatid becomes a separate chromosome. Then the cells divide again. There are now four cells, each with four chromosomes, one of each pair of chromosomes from the original. The process occurs in both egg and sperm formation. When the sperm fuses with an egg, each provides half the chromosomes for the new individual.

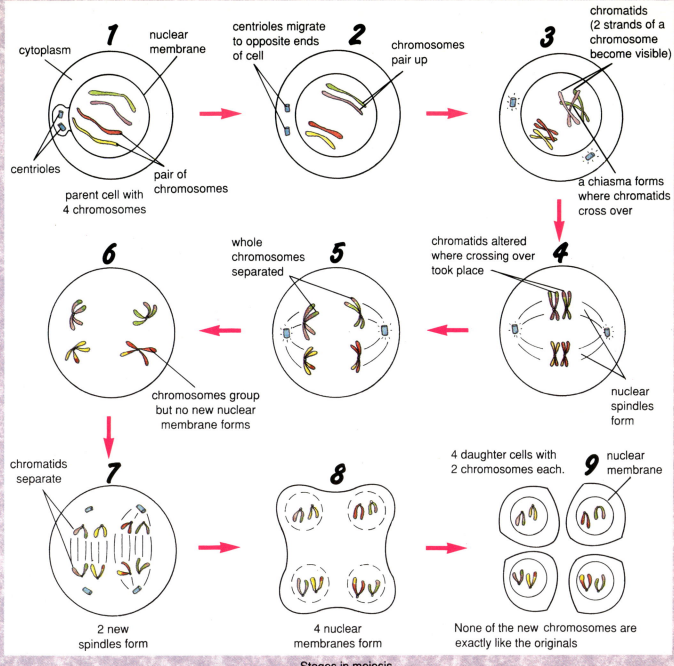

1. cytoplasm — nuclear membrane — centrioles — pair of chromosomes — parent cell with 4 chromosomes

2. centrioles migrate to opposite ends of cell — chromosomes pair up

3. chromatids (2 strands of a chromosome become visible) — a chiasma forms where chromatids cross over

4. chromatids altered where crossing over took place — nuclear spindles form

5. whole chromosomes separated

6. chromosomes group but no new nuclear membrane forms

7. chromatids separate — 2 new spindles form

8. 4 nuclear membranes form

9. 4 daughter cells with 2 chromosomes each. — nuclear membrane — None of the new chromosomes are exactly like the originals

Stages in meiosis

It is highly unlikely that there has previously ever existed a cell with this identical set of chromosomes, so each fertilized egg cell is **unique**. The advantage of sexual reproduction lies in the uniqueness of the fertilized egg that makes each new life just a little different from either of its parents. Variation among offspring may produce one or more that can adapt to changing conditions of the environment.

Once an egg has been fertilized, all subsequent cell divisions are mitotic and each new cell receives the full number of chromosomes.

Eventually the new organism will reach maturity, and its time will come to make a contribution to the survival of its kind. Its reproductive organs will then produce sperms or eggs and the cycle of life will have made its full turn.

Fertilization – the fusion of male and female pronuclei

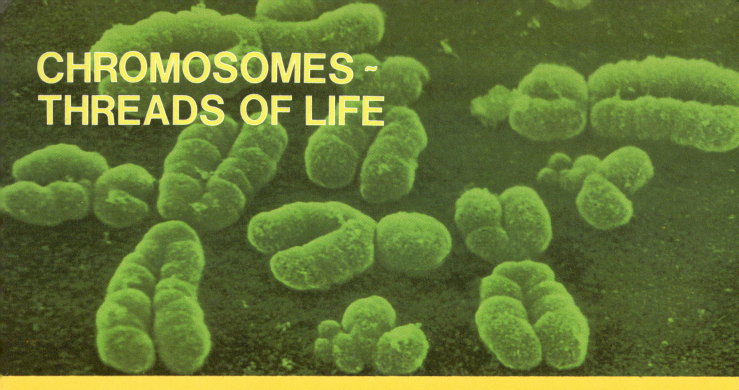

CHROMOSOMES ~ THREADS OF LIFE

With our present-day knowledge of chromosomes and meiosis we can explain Mendel's results, using a capital letter for a dominant gene and a lower case letter for a recessive gene. Thus **T** stands for the tall factor in peas and **t** represents the short factor. Although garden peas have seven pairs of chromosomes, we will work only with the pair that carries genes for size.

Think of two pea plants in flower, ready to produce sex cells. One is a tall plant. Both of its chromosomes carry the tall gene, **T**. The chromosomes of the other plant, a short one, each carry the short gene, **t**. Meiosis occurs and the chromosome pairs are divided by reduction division. The result is male sex cells with a chromosome bearing the **T** gene, and female sex cells each with a chromosome carrying the short gene, **t**.

When the cells join to form a fertilized egg, the reconstituted chromosome pair will have one **T** chromosome and one **t** chromosome. It will appear tall, because **T** dominates **t**. We can also see that if the original tall plant had been crossed with a tall plant carrying only tall genes, all the offspring would have **TT** chromosomes and would also grow tall.

A cross between the short plants would have similar results. That is why plants that always breed true to type are called **pure-bred**.

But let us go back to the plant **Tt**, with its mixed genes. When its cells go through meiosis it can form two kinds of sex cell, one with a chromosome carrying a tall gene and one carrying a short gene. When two plants of this type are crossed, we can get three kinds of fertilized eggs: **TT** when a tall gene sex cell meets another, **Tt** when a tall gene cell is paired with a short gene cell, and **tt** when two short gene cells fuse.

△ Chromosomes

This explains Mendel's findings. Remember that by crossing his second generation of peas he got some pure-bred tall plants (**TT**) and some short plants (**tt**), and some tall plants (**Tt**) that also produced both tall plants and short plants in the third generation.

Seven genes for seven chromosomes

1	Seed shape	round dominant to wrinkled
2	Seed colour	yellow dominant to green
3	Seed coat colour	coloured dominant to white
4	Pod shape	inflated dominant to wrinkled
5	Pod colour	green dominant to yellow
6	Flower position	axial dominant to terminal
7	Stem length	tall dominant to short

	t	t
T	Tt tall	Tt tall
T	Tt tall	Tt tall

	T	t
T	TT tall	Tt tall
t	tT tall	tt short

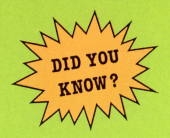
Mendel worked with seven different factors and found that they all combined and recombined in exactly the same way. Since it is now known that peas have seven pairs of chromosomes, we can see that each of the characteristics Mendel studied was located on a different chromosome pair.

Humans have 23 pairs of chromosomes and fruit flies have four pairs. Geneticists have studied dozens of fruit fly characteristics, and anyone knows that humans have more than 23 characteristics. There is only one way to explain this and it is to assume that each chromosome has many genes linked to it. One estimate is that each human chromosome contains at least 3000 genes, all of which contribute to producing our charateristics.

Some questions have to answered. We have a reasonable scientific explanation of why offspring resemble their parents and why black sheep sometimes produce white lambs. Why, however, are some of us male and others female?

In the 23 pairs of chromosomes there are 22 matching sets and an odd pair called **X** and **Y** (**sex chromosomes**). Actually, they look different; the human Y chromosome is short and round, while the X is rather long. The Y chromosome in fruit flies has a little hook, while the X chromosome is quite straight. It has been found that cells from male mammals have one X and one Y chromosome. Female cells have two X chromosomes. After reduction division, egg cells can contain only X chromosomes, but sperms can contain either an X or a Y. If an egg is fertilized by a Y-bearing sperm the offspring will be male. When two X-bearing sex cells unite, the offspring will be female.

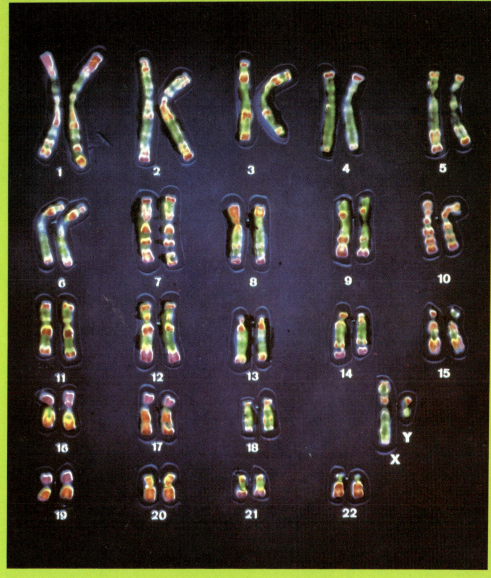

A human's set of chromosomes

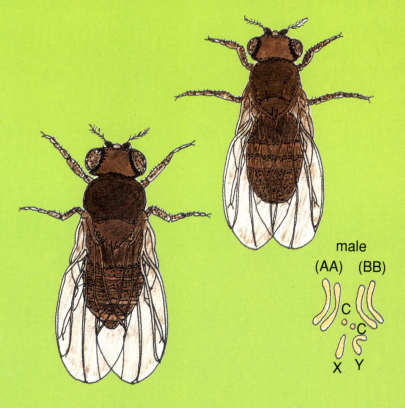

female
(AA) (BB)

male
(AA) (BB)

Fruit flies

CHROMOSOMES and HEREDITY

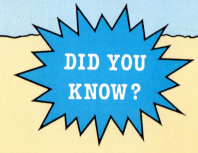

DID YOU KNOW?

JAPANESE MAY CHOOSE THEIR BABY'S SEX

The technique is a simple but effective one. It relies on separating in a **centrifuge** sperm that carry X chromosomes from those bearing Y chromosomes. Semen is placed in a viscous liquid called **parcoal** and spun in a centrifuge. Those sperms with X chromosomes separate towards the bottom of the centrifuge tubes.

Test tube babies are formed by *in vitro* fertilization in Petri dishes, and eggs are fertilized by sperms from the father. The researchers in the University of Tokyo say that the technique has been used to produce six babies in one year. They were all girls. They claim a 95% success rate when producing females.

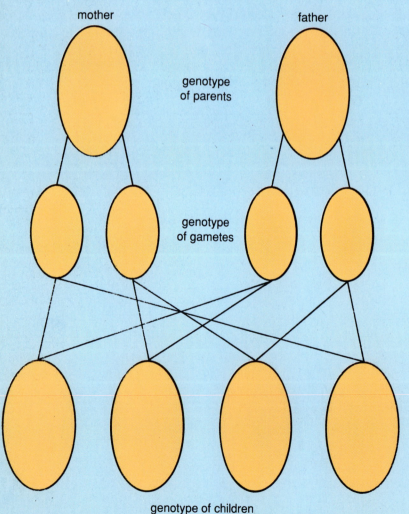

mother

father

genotype of parents

genotype of gametes

genotype of children

now try these!

C 1 Explain concisely how the Japanese technique could be used to control the sex of a baby.

2 a Copy the diagram opposite and fill in the sex chromosomes carried by the parents, gametes and children.

b Name the type of cell division that takes place in the formation of gametes.

c Name the type of cell division that takes place when ordinary body cells divide during growth.

d State one difference between the two types of cell division named above.

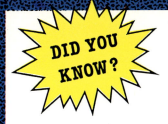

Chromosomes — Can Break —

It is now clear that genes are inherited independently of each other only when they are located on different chromosomes. If genes are linked together on the same chromosome they are inherited together. Occasionally, however, scientists find an individual in which genes that apparently linked in the organism's ancestors seem to have separated and relocated on some other chromosome.

In the early stages of meiosis the pairs of chromosomes twine around each other, and as they are pulled apart the threads may break where they are most twisted. In some cases, they join up again, but sometimes half of one thread joins with the thread from the partner chromosome. By this breakage and rejoining, called **cross over**, linked genes may be separated and new combinations of genes produced.

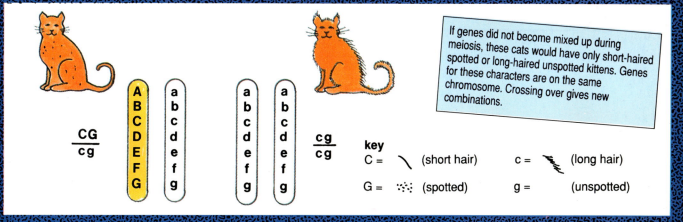

If genes did not become mixed up during meiosis, these cats would have only short-haired spotted or long-haired unspotted kittens. Genes for these characters are on the same chromosome. Crossing over gives new combinations.

$$\frac{CG}{cg} \qquad \frac{cg}{cg}$$

key

C = \ (short hair)   c = (long hair)

G = ∴ (spotted)   g = (unspotted)

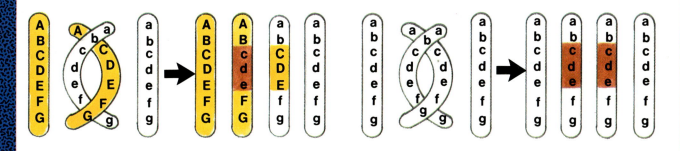

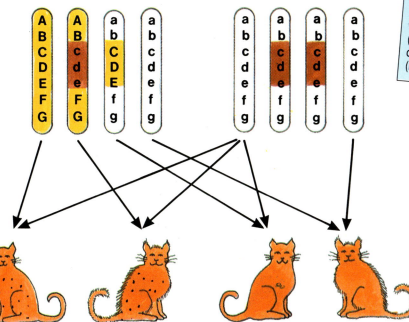

When chromosome pairs split at meiosis, the inner two of four coil around each other. On separating, the coiled two may break and join with part of the other. With **similar** chromosomes (right pair), crossing over does not alter chromosomes; with **dissimilar** chromosomes (left pair), it produces altered chromosomes.

A CLOSER LOOK AT CHROMOSOMES

By using special techniques, geneticists have been able to make **maps** of chromosomes to show the positions of genes. The genes that are easiest to locate in humans are those linked to the X chromosome; these are called **sex-linked**. One of the sex-linked genes in humans is the one controlling our ability to see colours. The gene that produces **colour blindness** is recessive to normal vision. Thus colour blind women must have the colour blind gene on **both** X chromosomes. Men can be colour blind with one abnormal gene, since there is nothing on the Y chromosome to correspond.

Sometimes an abnormal chromosome number can cause great misfortune. In humans many physical and mental abnormalities occur because of a failure of chromosomes to separate during cell division. A condition known as **Down's syndrome** in which the individuals are not only mentally and physically retarded but have facial abnormalities as well, is caused by a third chromosome in the number 21 set.

In some individuals there is only one X chromosome. These persons are undeveloped females with abnormal skin, skeleton and muscles. The chromosome pattern XXY produces underdeveloped and mentally retarded males.

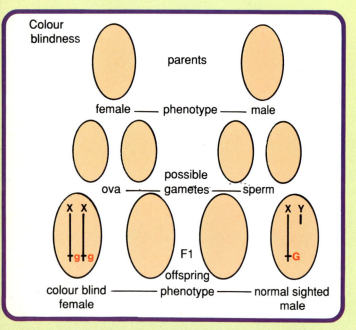

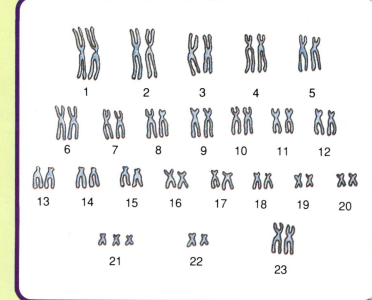

Set of human chromosomes of a person suffering from Down's syndrome

Occasionally something goes wrong during cell division; a cell with an abnormal number of chromosomes is produced. If the number of chromosomes is an exact multiple of the normal number (**diploid number**) such cells form a **polyploid** organism. Garden plants with double flowers are polyploids.

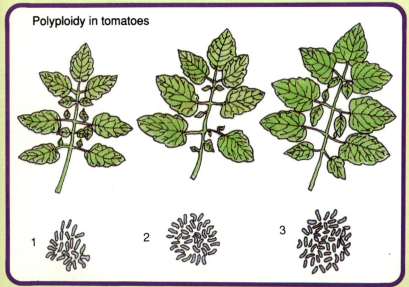

Polyploidy in tomatoes

NOW TRY THESE!

Red-green colour blindness is controlled by a sex-linked gene. There are two alleles, one for normal sight, represented by **G,** and one for the colour blind condition, represented by **g.** Copy the diagram (above left) and complete it to show the inheritance of this condition by inserting

a the genotype of the gametes

b the genotypes and phenotypes of the parents

c the genotypes of the other two possible offspring.

△ A Down's syndrome child

Tomato chromosome map

On the right is shown a **chromosome map** of one chromosome of a tomato. The letters show the positions of some of the **alleles**. Two or more genes are said to be alleles of each other when they occupy the same **relative** position on **homologous** chromosomes and when they both affect the same character. The alleles are arranged in a line along the chromosome. The dominant alleles are in capitals. The character controlled by the normal allele is illustrated on the left, and the character controlled by the mutant allele is shown on the right.

Most mutants are recessive, which is why they often do not show in the phenotype; they are usually harmful. This is another reason why their appearance is rare. A few are dominant and not fatal. The genes reading downwards are:

F	= germal fruit;	f	= fasciated fruit
A	= purple stem;	a	= green stem
HL	= hairy stem;	hl	= hairless stem
Lf	= normal branch;	lf	= leafy branch
J	= jointed branch;	j	= jointless branch
Nt	= normal fruit;	nt	= fruit with nipple
Wtl	= non-wilting;	wtl	= wilting

NOW TRY THESE!

True breeding, hairy-stemmed tomato plants when crossed with true breeding hairless-stemmed tomatoes only produce hairy-stemmed plants. When these hairy plants interbreed some plants with hairless stems are produced.

Explain fully how these results are obtained, setting out clearly the generations, phenotypes, genotypes and gametes.

Breeding

From the very earliest times, when humans first started farming and keeping animals, people chose the best seeds for each year's crops and the best animals for their needs. For example, some farmers wanted to keep beef cattle and others wanted dairy herds. Then again, another farmer might have windswept fields and would want a short, sturdy cereal, whereas another, with sheltered fields, might prefer tall grain.

Gradually over thousands of years, hundreds of different kinds of domestic animals and cultivated plants have been developed. Once-wild plants changed almost beyond recognition. Cereals, for instance, were developed from wild grasses. Cauliflower, broccoli, cabbage and Brussels sprouts all belong to the same family, which came from the wild plants of the seashore. Similarly, all the varieties of chicken developed from the jungle fowl. The ancestor of the pig is the wild boar, that of the cow the wild ox, that of the dog the wolf, and so on.

Even though farmers had been changing and improving the quality of their stock for thousands of years, the science of **genetics** did not develop until the pioneering work of Gregor Mendel.

The main method by which organisms are improved is **selection**. This means that the breeder chooses only plants or animals with good qualities to be the parents of the next generation.

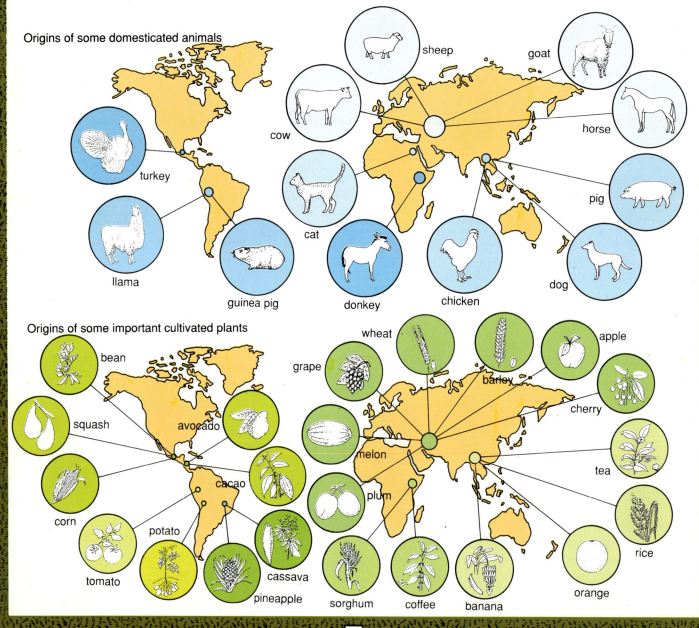

Origins of some domesticated animals

sheep · goat · cow · horse · turkey · pig · cat · llama · dog · guinea pig · donkey · chicken

Origins of some important cultivated plants

bean · wheat · apple · grape · barley · cherry · squash · avocado · melon · tea · cacao · plum · corn · potato · cassava · rice · tomato · pineapple · sorghum · coffee · banana · orange

If he wants to improve his cattle, he mates a good cow and bull. If he wants to improve his beef cattle even more he may mate a brother and a sister. This concentrates the quality and is known as **inbreeding**.

Another important method in breeding is that of **hybridization**. This means that two plants or animals of different varieties are used for breeding. The resulting **crossbreed** or **hybrid** is often stronger and more vigorous than either of the parents. Modern types of corn are an important example of hybridization. They are stronger and produce more nourishing grain than their ancestors did.

Animal breeding

Any animal breeder will have in mind an ideal type of animal for any one purpose. Dairy cows are judged on the quality and quantity of milk they produce. Sheep are judged on the qualities of their wool, and so on. There are about 200 different breeds of sheep, such as the **Merino** for wool and the **Down** for the meat. There are 300 breeds of pig, of which there are three types – **lard**, **bacon** and **pork**. Lard pigs are large and must have a lot of fattening food such as corn. Bacon pigs are smaller with long bodies and pork pigs have long lean bodies.

It is of course important to give a breed the most suitable conditions in which to live. Even if a cow has some very good genes for the quality of producing milk, it will not do so unless it has enough good, green grass.

Another technique is **artificial insemination**. This means that two animals do not mate in the normal way. The sperm is taken from a selected male and injected into the reproductive canal of the female. In this way one male can act as the father to hundreds of offspring with selected qualities.

The technique of 'test tube baby production' or **in vitro fertilization** goes a step further. The eggs from a selected female can be fertilized by the sperm of a selected male, in a Petri dish. The **zygote** produced into externally can now be introduced into the **uterus** of a female animal of the same species. In this way, young embryos can be frozen and transported anywhere in the world to improve the quality of stock.

a   Lard pig

b   Bacon pig

c   Wild pig

e *Now try these!*

Domestic sheep differ from their wild relatives because their coats are made mainly of soft wool and lack the straight hair that normally hides the winter wool of the wild forms.

Over 50 separate breeds of sheep are now recognized in the British Isles. The East Friesian sheep are renowned for their milking capacity. The short wool sheep grow quickly and produce good meat. Shetland sheep have soft fine wool that is used to knit shawls and warm clothing .

Use the information in the above passage to explain the meaning of the following terms.

a   variation
b   artificial selection

PLANT BREEDING

The same basic principles apply in plant breeding as in animal breeding. The breeder selects desirable qualities in the particular plant. **Pollen** is then taken from the **anthers** and dusted on to the **stigma** of another flower. Only the pollen chosen must fertilize the plant. Therefore the plants must be protected by plastic bags so that pollen carried by the wind or insects will not interfere with the results.

Hybridization is used a great deal in food plants such as tomatoes, onions and corn. Modern hybrid corn has a yield 20 times as great as that of its ancestors.

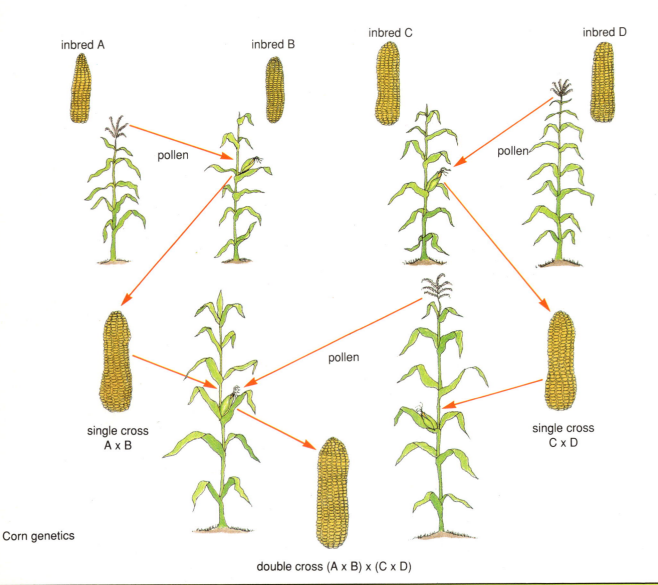

inbred A   inbred B   inbred C   inbred D

pollen   pollen

pollen

single cross
A x B

single cross
C x D

Corn genetics

double cross (A x B) x (C x D)

NOW TRY THESE!

1 Disease resistance in wheat is sometimes controlled by a recessive gene. If a homozygous normal plant is crossed with a disease-resistant plant, the F1 will contain the gene for resistance. Explain how these F1 plants can be used to produce a disease-resistant crop. In your answer, explain the genetical details for
   i   producing the F1
   ii   producing the F2
   iii producing 100% disease-resistant plants.

2 When the original normal plant is crossed with a disease-resistant plant, the problem is to prevent self-pollination. Biotechnologists have developed a chemical which can be sprayed on young wheat plants to prevent pollen production.
   a Explain what is meant by self-pollination.
   b Name the parts of the flower which are affected by the chemical spray.
   c Explain why self-pollination must be prevented when trying to produce disease-resistant crops.

MINI BULLETS
—HELP—
PLANT BREEDERS

A new method of putting genes into cells will help scientists to add genetical material to some types of plants where it was previously thought to be impossible.

Geneticists can improve the yield of crops and make plants larger and more resistant to disease by inserting new genes into existing species. At present, they have to remove the tough cell wall before they can insert genetical material into the cell's nucleus. Another method is to bore a hole through the wall of a cell and pass the genetical material through it.

American scientists have designed a new technique. They prepare the genetical material for transfer into the cells in a well-known way (see page 28) i.e. they put it in circular microbial **plasmids**. These are then shot into the **host cells**.

First the plasmids are mixed with tiny particles of tungsten. These are then stuck on the front of a cylindrical plastic bullet. The device works like a miniature pistol. A firing pin detonates a blank gunpowder charge that propels the bullet down a barrel onto a plate. The impact of the bullet hitting the plate jerks the tungsten and plasmids off the bullet's surface through an opening, across a vacuum, to the plant cells. The particles of tungsten are large enough to penetrate enough cells to be effective but they do not destroy the cells.

The researchers tested their idea by inserting genes that were known to make a particular enzyme. After one bombardment, the cells produced 200 times the normal amount of enzyme!

NOW TRY THESE!

C

1 What is meant by 'genetical material'?

2 Suggest a disease to which the crop could be made resistant.

3 Describe how you might produce a crop capable of the fixation of nitrogen, based on your knowledge of leguminous plants in the nitrogen cycle and genetic engineering.

Genetics and Society

We are not amused!

Haemophilia – a royal complaint

Haemophilia is an hereditary disorder of the blood. It is transmitted by an abnormal X chromosome. It prevents the blood from clotting properly when it comes in contact with air. A girl will not be directly affected by haemophilia since she always has two X chromosomes and one of these will almost invariably be normal. The normal X chromosome overrules the effects of the abnormal one but she is still able to transmit the problem to some of her children. When an abnormal X chromosome meets a Y chromosome, the result is a boy with haemophilia. His sons will not suffer from it because they will inherit a normal Y chromosome but his daughters may transmit the disease because of the defective X chromosome they have received from him.

DID YOU KNOW?

Haemophilia helped shape the history of Europe because it involved members of British royalty. Queen Victoria passed it on to the reigning families of Spain and Russia at the beginning of the twentieth century.

Because of her son's haemophilia, the Tsarina of Russia allowed Rasputin to influence his country's foreign and internal policies. The Tsarina thought that Rasputin had magical powers that could cure her son's problem.

Today, when haemophiliacs need an operation of any sort, they are first given a plasma extract containing a concentration of anti-haemophilia substances. Despite the availability of this treatment, haemophiliac people must be protected throughout their lives from blows and injuries. They are usually advised to find an indoor profession, for example, an office job.

Looking at cells

It is possible to tell the sex of a foetus or to check for genetic defects such as Down's syndrome with a technique called **amniocentesis**. A sample of amniotic fluid from a pregnant woman is taken, as shown in the photo.

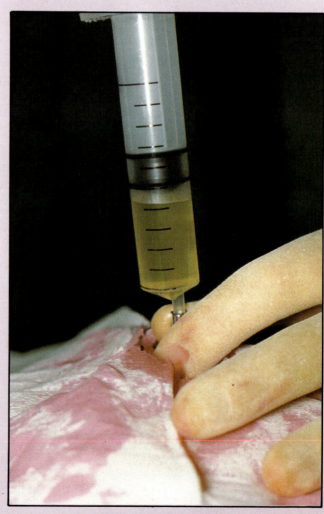

△ Amniocentesis in progress

Some cells from the foetus will be floating in the fluid and can be collected and examined with a microscope.

—Now— try these!

c

1 What is the function of the amniotic fluid surrounding the foetus?

2 What difference would you expect to find between a male cell and a female cell?

3 What part of the cell would you need to examine?

Cells Give Us a Clue

DID YOU KNOW?

Doctors sometimes need to look at a person's chromosomes to see if the chromosomes have something wrong with them. To do this, body cells are grown in a liquid containing oxygen. The cells are treated to make the chromosomes clearer and then placed in a dilute solution. Water moves into them and this makes them swell up. The cells are then placed on a slide, squashed and stained. After staining they are examined under a microscope.

▽ Down's karyotype

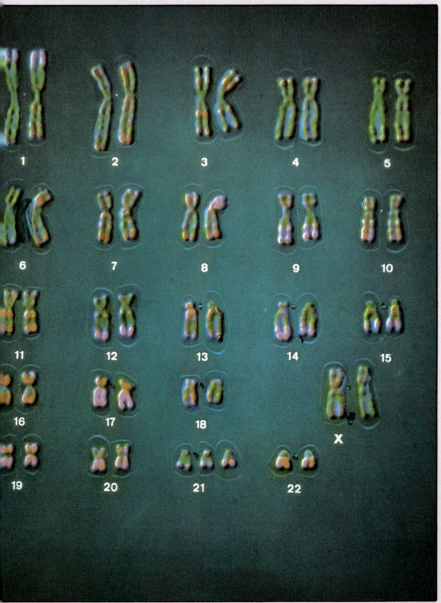

—Now— try these!

e

1 Explain why the cells need oxygen in the liquid.

2 What is meant by a dilute solution?

3 Name the process by which the water moves into the cells.

4 Explain why the water moves into the cells.

5 Suggest why the cells need to be stained.

6 The method described can show us what the chromosomes in a single cell look like. The photograph opposite shows the chromosomes taken from a person suffering from Down's syndrome.

   a Which group of chromosomes tells you that they are from someone who has Down's syndrome?

   b The chromosomes in group 23 are the sex chromosomes. What is the sex of this person? Explain how you can tell.

GENETIC ENGINEERING

Genetic engineering in a nutshell

Genetic engineering means isolating a gene from one organism and putting it in another. In biotechnology scientists often isolate the human genes that make human cells produce, for example, **hormones** (e.g. insulin) or blood proteins. They insert these genes into bacteria or yeasts.

The idea of transferring the human gene to the bacteria is to increase production. The microbes grow relatively cheaply in large fermenters, providing almost unlimited amounts of substances that are practically unobtainable in bulk in any other way.

Genetic engineering starts with the biologist using biological 'scissors' called **restriction enzymes**. These cut chains of **deoxyribose nucleic acid (DNA)** at specific points. The DNA chains are present in chromosomes. With the restriction enzymes the biologist carves out very precisely the gene that he or she wants from the hundreds of others in the nucleus of the cell. The next stage is to insert the gene into a bacterium. It is not put directly into the bacterial chromosome. Instead, genetic engineers use a circular piece of DNA called a **plasmid**. These are normally present in bacteria but are largely independent of the rest of the cell. Plasmids, like chromosomes, carry bacterial genes which control the microbe's metabolism.

The plasmid is cut open with restriction enzymes and the foreign gene inserted. The break is sealed with another enzyme called a **ligase** (an enzyme which binds chemicals together). This process creates a **hybrid** (mixed) molecule called **recombinant DNA**.

The altered (**infective**) plasmids are then mixed in a test tube with bacteria which do not have any plasmids. Some plasmids move inside these bacteria. The infective plasmids carry the foreign gene inside the cell where it can instruct it to make the required protein. The cells are then called **mutants**. The process is shown in the diagram opposite. The experiment is usually regarded as a success only when the gene is **expressed**. This means that the host cell obeys the instructions carried by the foreign gene and manufactures human proteins.

How genetic engineering saves lives

Insulin production

Insulin is a chemical called a **hormone** or 'chemical messenger'. Its function is to keep the concentration of glucose in the blood at a constant level (0.1 g per 100 cm^3 of blood). If the glucose level falls much below this, the body does not have enough glucose to release energy. A glucose level above 0.1 g per 100 cm^3 also disturbs body functions. In particular, the ability of the kidneys to reabsorb glucose is lost and glucose appears in the urine. The body is eventually drained of fuel.

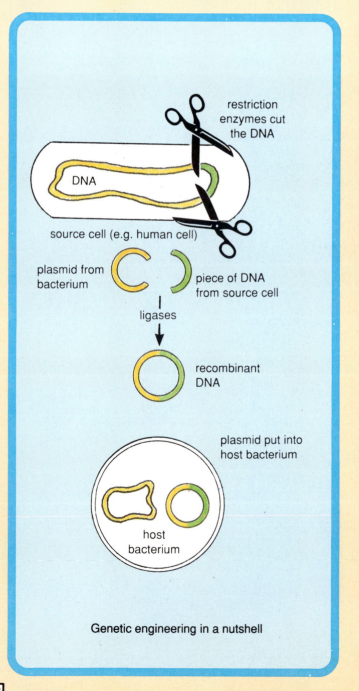

Genetic engineering in a nutshell

A gland in the abdomen, called the **pancreas**, normally makes insulin. It helps to keep the concentration of glucose constant by changing excess glucose into an insoluble carbohydrate called **glycogen**. This is stored in the liver. Without enough insulin people cannot control their glucose level and suffer from **diabetes**. Fortunately, it is possible to control this condition by injecting pure insulin into a diabetic person.

Until recently the insulin used for this treatment was taken from the pancreas of pigs and cattle. It was difficult to produce enough insulin this way. Genetic engineers have solved the problem of mass production of insulin using bacteria. The method is summarized in the flow diagram opposite.

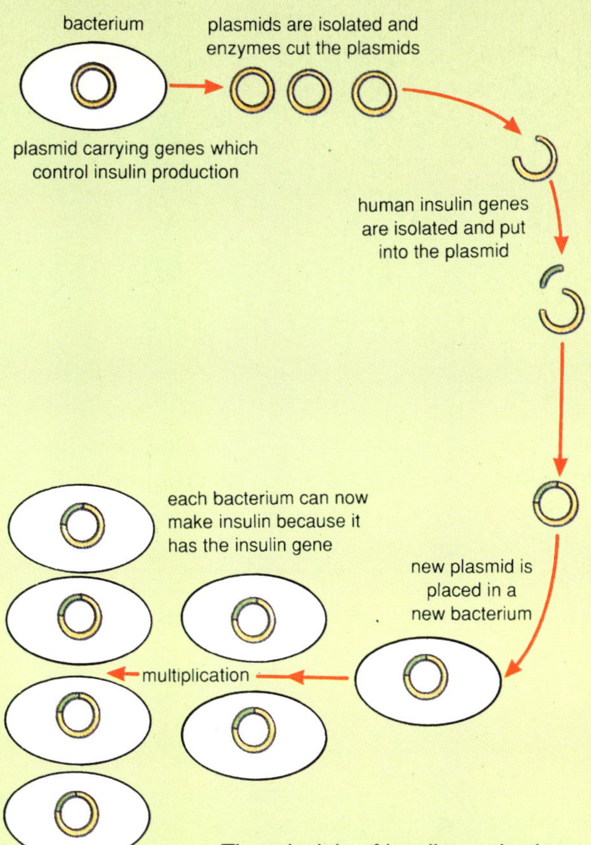

bacterium

plasmids are isolated and enzymes cut the plasmids

plasmid carrying genes which control insulin production

human insulin genes are isolated and put into the plasmid

each bacterium can now make insulin because it has the insulin gene

new plasmid is placed in a new bacterium

multiplication

The principle of insulin production

Do-it-yourself Pesticides

DID YOU KNOW?

Biologists have grown a tobacco plant that makes its own pesticide. A Belgian genetic engineering company has modified the genes of a tobacco plant so that it produces a toxin that kills insects.

The gene which controls the manufacture of the toxin has been isolated from a bacterium. It has been inserted into the tobacco plant in two ways:

* The cellulose cell walls of some of the leaf cells were broken down so that the plasmid could pass freely into the cytoplasm.

* Fragments of the stem were wounded with preparations containing the genetically altered plasmids.

Mutants in Australia

DID YOU KNOW?

The releasing of genetically engineered microorganisms into the environment to control plant production is about to spread to Australia. A scientist at the Waite Agricultural Research Institute in Adelaide received final clearance from the university to do his research. He wants to soak the roots of almond seedlings in a solution of genetically altered bacteria to treat crown gall disease in roots. After completing the four-month experiment, he will sterilize and dispose of the affected soil.

Now try these!

e 1 Suggest how the cellulose cell walls of the tobacco cells might have been broken down.

2 Do you think there are possible dangers to the environment involved in this technique?

Now try these!

e 1 Explain why the scientist sterilized the soil before disposing of it.

2 Suggest how the sterilization could be carried out.

3 Discuss why environmentalists might oppose the introduction of mutant bacteria into the environment.

4 Suggest how the bacteria would be able to combat crown gall disease (a disease caused by a bacterium) in plant roots.

Some People are Special

Mistakes sometimes occur during the setting up of the **chromosome pattern** for a new individual. These mistakes can often be seen by examining the chromosomes. **Chromosome analysis** may be started with the growing of a culture of white blood cells in a glass tube filled with nutrient and kept at the correct temperature. Cell division is stopped by the addition of a special chemical (**colchicine**). Cells are placed in a salt solution which causes expansion and separation of the chromosomes. The culture is then spread on a glass slide, stained, and placed under a microscope. A camera attachment photographs a single cell with its chromosomes.

An enlarged copy of the photograph serves as a reference. From another copy, the silhouettes of the individual chromosomes in the cell are cut out to be assembled by hand, in pairs according to shape and size, on a prepared sheet. The last step in this procedure consists of fixing the shapes on the paper. The end product is called a **karyogram** and is photographed to make a permanent record.

Preparation of a karyogram

1 A blood disorder leading to a form of anaemia has been found to be inherited. Investigation of the medical records of a particular family yielded the following information, though in some cases the records gave no information regarding the presence or absence of the symptoms.

a What is meant by the term 'carrier'?

In your answers to each of the following questions explain your reasoning clearly.

b What is the likely genotype of Alan?

c What is the probability that Clare is a carrier?

d If Frank was to have children with a woman with the same genotype as Sarah, what proportion of any sons might be expected to be anaemic?

2 Some people can roll their tongues into a U-shape.

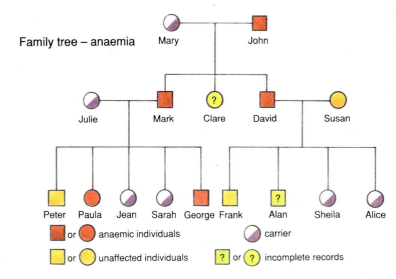

Family tree – anaemia

■ or ● anaemic individuals ◖ carrier

■ or ● unaffected individuals ? or ? incomplete records

The tongue-rolling characteristic

Tongue-rolling is controlled by a dominant gene **R**. Of three children in a family, two were tongue-rollers, the other was not. Of the parents, one was a tongue-roller, the other was not.

a By means of a diagram, show the genetic make-up of this family in respect of tongue-rolling.

b Three of the grandparents were not tongue-rollers.

  i What was the phenotype of the fourth grandparent?

  ii What were the possible genotypes of the fourth grandparent?

3 Huntington's chorea is a rare but very serious inherited disease caused by a dominant gene, **H**. The effects of the disease do not appear until the age of about 30 years.

, A woman of 25 is planning to start a family, but the woman's father has Huntington's chorea and is heterozygous for the condition. There is no history of the disease in her mother's family.

a What are the chances of the woman having Huntington's chorea?

b What advice might be given to the woman by a genetic counsellor about the desirability of starting a family?

c By means of a diagram, explain the reasoning on which the advice is based.

4 Humans sometimes grow up with bones that are brittle and easily break. This condition is passed on from parents to their children, as shown in the family tree below.

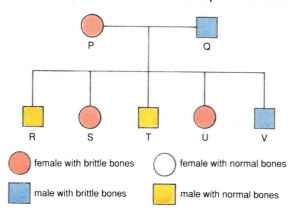

● female with brittle bones ○ female with normal bones

■ male with brittle bones ■ male with normal bones

Inheritance of brittle bones

a What is the sex of the child labelled U?

b Does the child have brittle bones?

c Which is recessive, normal or brittle bones?

d Are the two parents homozygous or heterozygous?

e The gene for brittle bones is **B**. The gene for normal bones is **b**. What are the possible genotypes for children S and T?

Allele   One of the alternative forms in which a gene may exist

Chromatids   The two parts of a chromosome when the chromosome is seen to be split in half lengthways at cell division

Chromosomes   The thread-like structures present in the nuclei of cells composed of genes along their lengths

Codominance   Two or more alleles making a positive contribution to a phenotype; one balances out the other and results in blending of characters

Contrasted Characters   Pairs of characters, one of which is dominant, the other is recessive

Dihybrid Inheritance   The inheritance of two pairs of contrasted characters

Diploid   A cell or organism with two sets of chromosomes, each chromosome having a partner

Dominant Factor   One of a pair of genes which expresses itself in the heterozygote

Gamete   The parental contributions in sexual reproduction, which fuse to form a zygote

Gene   A subdivision of the genetic material responsible for determining the structure of a particular protein or chain of amino acids. It represents Mendel's 'germinal unit' and determines all hereditable characters

Genotype   The genetic constitution of an organism

Haploid   The possession of only one set of chromosomes

Heterozygote   An individual receiving unlike genes from both parents

Homozygote   An individual receiving similar genes from both parents

Hybrid   Used by Mendel to mean any offspring produced by sexual reproduction between two visibly different parents; usually taken to mean an offspring of two parents of different species

Incomplete Dominance   The presence of one of a pair of alleles which does not exert the full effect; this results in blending of characters; the full effect is only seen when both alleles are present in a homozygote

Meiosis   The type of cell division in which the chromosome number is halved, so that the diploid cell gives rise to haploid cells (often gametes)

Mitosis   Cell division in which daughter cells have the same number of chromosomes as the parent cell

Monohybrid Inheritance   The inheritance of one pair of contrasted characters

Mutation   Any change in the genetic material; it may result in the conversion of one allele of a gene to an alternative allele

Phenotype   The outward expression of genes

Recessive Gene   One of a pair of genes which does not show itself in the heterozygote but can be passed on, unaltered, to future generations and will show itself in the homozygote

Selection   The process by which organisms of different genotypes leave different numbers of descendants, so that the gene frequency of the population is changed in future generations

Species   A group of organisms, the members of which appear similar, and which interbreed to produce fertile offspring